The Age of Castles

HOW CASTLES WERE BUILT

Peter Hicks

PowerKiDS press.
New York

Published in 2008 by The Rosen Publishing Group, Inc.
29 East 21st Street, New York, NY 10010

First Edition

Series editor: Alex Woolf
Editor: Jonathan Ingoldby
Designer: Peter Laws
Cover Designers: Rachel Hamdi and Holly Fulbrook
Illustrator: Peter Dennis
Project artwork: John Yates
Consultant: Richard Eales, Senior Lecturer in History, University of Kent.

Library of Congress Cataloging-in-Publication Data

Hicks, Peter, 1952-
How castles were built / Peter Hicks. — 1st ed.
p. cm. — (The age of castles)
Includes index.
ISBN 978-1-4042-4293-7 (library binding)
1. Fortification—Europe—Juvenile literature. 2. Castles—Europe—Juvenile literature. I. Title.

UG428.H532 2008
623'.194—dc22

2007032589

Manufactured in China

Picture Credits

The publishers would like to thank the following for permission to publish their pictures: (t=top; c=center; b=bottom; l=left; r=right) AKG 25tr; Ancient Art and Architecture Collection 32l, 38l; The Bridgeman Art Library, London /British Library, London 8bc,/Victoria & Albert Museum, London 9tl, /British Library, London 10lc, 10rc, /British Library, London 16c, /John Bethell 19c, /Bibliothèque Nationale, Paris 37tr; C.M. Dixon 42b; e.t. archive 7tr, 7bl, 26l, 26r; Eye Ubiquitous 17tr, 33tl; Sonia Halliday 21tr, 25bl; Robert Harding 5tr, 11t, 20c, 22l, 28c, 40c; Peter Hicks 15l, 15r, 17b, 18c, 35bc; Michael Holford cover, 6c, 9rc, 13t, 14c, 35tl, 36r; Angelo Hornak 29tl, 41c.

CONTENTS

What was a Castle?

A discussion is taking place. While his men-at-arms wait nearby, the baron listens to the advice of his engineers, who suggest the best place to site his castle. The baron listens patiently as he carefully studies the surrounding land. He decides. It will be on the low ridge above the river, on the edge of the busy market town.

What was a medieval castle? Basically, it was someone's house, specially strengthened against attack. Its owner was usually a powerful landowner, a baron, although the king owned many castles, too. They were built to protect the barons' large farming estates, which made these men very wealthy. A castle contained soldiers, or men-at-arms, and was their base, providing shelter, food, and weapons for the men and their warhorses. Along with these fighting men, the castle housed the baron's family and all his servants, so they were busy, bustling places when the baron was there.

▷ Many people consider Beaumaris on Anglesey the perfect castle. With its superb arrangement of moat, walls, turrets, and gates, entry for unwanted guests was almost impossible.

Because castles had to withstand possible attacks, they were strengthened in many clever ways. Deep ditches were dug around them and if they contained water, they were known as *moats*. High walls were built to make it difficult for attackers to climb in, and as castles developed, large towers, or turrets, were added to allow the defenders to fire arrows at them.

SITING A CASTLE

We often think of castles perched high up on rocky hills or mountains—excellent lookouts and very difficult to attack. These castles do exist, such as Manzaneres Castle in Spain, with its walls, turrets, and battlements. However, the vast majority of castles were built on low ground. What made the engineer-builders site their castles where they did?

△ At Manzaneres Castle in Spain, the tall turrets and walls allowed defenders a good field of fire to pick off attackers.

The most important reason was usually water supply. Nearby rivers and streams were no good (although they could be diverted to fill the moat), because they could be dammed, or worse, poisoned by an attacking army. It was essential for the water supply to be within the castle itself, so deep wells were dug into the water table. Next came ease of access. A castle was a very important part of the community, and people needed to be able to visit regularly and without difficulty.

For example, many castles doubled up as farms, so it was important that both workers and animals could get into nearby fields easily. Food had to be carried back into the castle for storage as this description of 1274 shows. The castle had: "a pantry and buttery ... a stable for sixty horses ... a barn, a cowhouse, a stable for cart and draught beasts, a bakery, and buildings for calves and pigs." Finally, since it was a military base, cavalry had to be able to leave the castle swiftly and move into the surrounding countryside.

▷ A castle moat meant that siege towers like this one—packed with infantry and archers—had trouble getting close enough to the walls.

Why Lowland Castles?

Castles were usually situated close to where people lived, and in medieval times this tended to be in low river valleys. Being overlooked by high ground was unimportant. Crossbow bolts, even those fired from a large "ballista" like this one, only had a range of 330 yards (300 meters), and catapults could only hurl stones about 165 yards (150 meters). Damp valley clay protected castles against tunneling and allowed moat defenses.

Royal Castles

THE ROYAL PROCESSION approaches one of the king's castles, and after several days traveling, the king is looking forward to the comfort the castle will offer. He has been visiting his castles in the south, consulting with his barons and making sure the coast is properly defended.

The kings of France and England were obsessed by castles. They built them, inherited them, and even stole them from their barons. Why were royal castles so important? First, kings needed accommodation when they traveled around their kingdoms. Medieval kings constantly moved around. Second, a number of castles were essential to England's defense. Because southeast England was a likely place an invader would land, the castles in Dover, Canterbury, and Rochester, and the Tower of London were known as the "keys of the kingdom."

△ King Richard I and his barons. In his first year as king, Richard spent £3,500 on Dover Castle and the Tower of London alone. At this time, the king's normal income from all of England was only £25–30,000 a year.

The Feudal System

When the Normans conquered England, they brought a method of control—the feudal system. The king gave land to his supporters, the barons. In return, they had to swear loyalty to him and provide soldiers for him. Barons were allowed castles in which they collected rents, kept records, held trials, and jailed wrongdoers. Below the barons were the knights who also received land if they promised to fight when they were asked. Below them came the peasants, who worked for everyone!

▽ Chateau Gaillard became Richard's favorite castle during the last two years of his life. Unsurprisingly, he called it the "fair castle of the rock."

Richard I, the "Lionheart," ruled over parts of France as well as England and was determined to defend the borders of Normandy. The result was Chateau Gaillard, built in less than two years! Richard was incredibly proud of his castle and when he visited it in 1198, he exclaimed: "Behold, how is this year-old daughter of mine?" Later he boasted: "I could hold this castle even if its walls were made of butter!"

ROYAL APPROVAL

THE BARON ENTERS the great hall. At the far end sits King John, splendid in robes and crown. The visitor bows down as the king begins a speech praising his loyalty and assistance over the years, and concludes by saying that the baron's request to build a new castle is granted. The baron has the royal approval he had been hoping for.

△ King John, shown here on his "Great Seal." During his reign (1199–1216), he approved the building of Kirkoswald Castle in Cumbria, England, among others.

△ Henry II (left) and Henry III (right) both suffered from baronial revolts during their reigns. In the fighting, they had to attack and even dismantle castles that belonged to troublesome barons.

The king was not interested in every single castle in his kingdom. Many were small with no military importance. However, he had to keep an eye on the larger, stronger castles that belonged to the most powerful landowners. This was because they contained many soldiers and huge stores of food and weapons that could be turned against the king. He was eager to see existing or newly-built castles in the hands of friendly barons.

"License to Crenellate"

When a landowner decided to seek the king's approval to build a castle or convert his house into one, a "license to crenellate" was sometimes granted. In 1281, Edward I allowed:

"Our beloved and faithful Stephen of Penchester and Margaret his wife to fortify and crenellate their home at Allington in the county of Kent with a wall of stone and lime, and that they and their heirs may hold it for ever. Witnessed by myself at Westminster on the twenty-third day of May in the ninth year of our reign."

MOTTE AND BAILEY CASTLES

THE HUGE EARTH AND TIMBER CASTLE in Lincoln takes shape. To make way for it, 160 houses in the city have been demolished, and harsh Norman soldiers force hundreds of local men to dig the earthworks and drag the timber from the woods nearby.

When an army conquered another region or country, the local people were often hostile. The victorious troops urgently needed to build safe bases for themselves. The quickest and easiest way was to build castles out of earth and wood. These materials were plentiful and cheap. The earth was simply dug, and woodland covered huge areas of the countryside. These early strongholds were called *motte and bailey* castles.

◁ Part of the Bayeux Tapestry showing Hastings Castle under construction, with English laborers building the motte.

The tall mound, or "motte," was built out of the earth dug from a large circular ditch. It was piled up in the center and packed down solid by the laborers. The top was flattened, which acted as a platform for a timber wall around the outside edge. Whole or split timbers were slotted in to postholes in the earth and their tops were sharpened. Inside, a tower was built to house the baron and his family. One in Ardres in France was three stories high and was: "a marvellous example of the carpenter's art ... piling storeroom upon storeroom, chamber upon chamber ... larders and granaries and a chapel."

Forced Labor

The large number of motte and bailey castles built in the eleventh century meant that thousands of laborers were required for their construction. There is evidence that local people were forced to work for the Normans. The Anglo-Saxon Chronicle tells us that they:

"built castles far and wide throughout the land, distressing the wretched people; and ever after greatly grew in evil."

This points to another reason why these castles were built so fast—there was very little need for skilled labor.

DEFENSES

The motte was a refuge where defenders could retreat and continue fighting if the main castle was overrun. The motte was very steep, which made it difficult for attackers to climb. To make it even harder, the motte was sometimes covered with wooden boards, which many attackers found themselves sliding down! The boards also helped to stop soil in the motte from washing away.

△ The view across the moat and bailey bank to the motte of Berkhampstead Castle. No wooden features remain, because as you can see, they were replaced by stone.

The second part of the castle was the large enclosed courtyard, or "bailey." This was protected by a deep ditch or moat and a high earth bank topped with a bound timber fence—known as a *palisade*—running its entire length. There was usually one entrance to the bailey, over a bridge across the ditch or moat, and in through a high wooden gateway.

The Link Between the Motte and the Bailey

Access to the top of the motte was usually provided by a wooden bridge. The Bishop of Therouanne in France described one: "A bridge, leading from inside the ditch, was gradually raised ... supported by sets of piers so as to reach the level of the mound, landing at its edge." A similar bridge can be seen in this detail from the Bayeux Tapestry.

The bailey contained all the buildings for everyday life—the kitchens, a hall for eating and sleeping, an iron-smith's forge, stables, and a chapel. In Stansted Mountfitchet, one of these early castles has been reconstructed, and part of it is shown below. Notice the wooden palisade fence and the tower (built on stilts) for a lookout. This castle does not have a motte, just circular ditches and banks, called a *ringwork*.

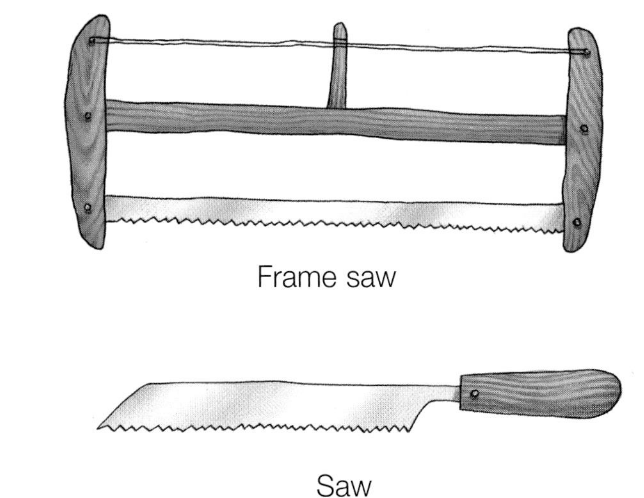

Frame saw

Carpenter's axe

Saw

EARLY STONE CASTLES

CURTAIN WALLS

A LONG LINE OF HORSES AND CARTS are slowly walking toward the castle. They carry tons of rough flint from the surrounding chalk hills. Each cart dumps its load at the base of the bailey and heads back for the quarry. An army of men carry the sharp stones to the top of the bank—a "curtain wall" is being built.

By the early twelfth century, a large number of earth and timber castles were replaced by stone. Timber lasts for about 20 years, but it had two major disadvantages: it rots, and an attacking army could easily set fire to it. When the earth banks had hardened, it was safe to replace the timber walls with stone. The bailey was enclosed by a continuous curtain wall.

△ The solid stone wall defenses being built at Marseilles in southern France. Note the gatehouse guarding the bridge.

▷ The impressive gateway of Caerlaverock Castle in Scotland. It was strengthened by dozens of "machicolations"—slots through which missiles could be dropped on the heads of attackers.

The weakest part of a castle was always the entrance. Castle builders worked hard to make sure gateways were as secure as possible. Early stone gateways were simple stone square boxes, with an inner and outer entrance blocked by thick oak doors. This was improved upon with huge towers to guard both sides of the entrance.

▷ The wallwalk at Portchester Castle. The builders were so impressed with the walls of the Roman fort, they sited their castle inside. The three turrets are actually Roman.

It was essential that defenders could fight from the curtain walls. To make this possible, many were equipped with wallwalks—protected walkways, so the soldiers could be quickly sent to any part of the wall under attack. The walk was protected by a higher wall called a *parapet*. Gaps were left, called *embrasures*, which the soldiers could fire through. The solid parts were called *merlons* and soldiers used them to take cover.

SHELL KEEPS

It wasn't only timber walls that disappeared—wooden towers were also ripped down and replaced by stone. An expensive way of strengthening the motte was by encasing it with stone, but this was rare. Most castle engineers simply built a stone circular wall, 6½–10 ft. (2–3 m) thick, around the lip of the motte. Because the towers were hollow, they were called "shell keeps."

△ Arundel Castle in Sussex, England, is a dramatic and powerful example of a shell keep. It was in such an important position (by a river, near the coast) that the king often "moved in" whether the baron liked it or not.

Entry to the shell keep varied. Timber bridges were sometimes replaced by stone, although this required space. Often engineers laid stone steps into the side of the motte. At the top, the entrance was usually strengthened by a small arched gateway, with thick oak doors. To complete the defense of the castle, curtain walls were sometimes extended up the sides of the motte to link up with the keep—quite an engineering feat.

△ Even though space was limited, the interiors of early castles could be comfortable.

Obviously, there was limited space inside the shell keep, so how was it used? A hall was built for eating and meetings, with the fireplace and chimney set into the inside wall of the shell. Next to this was the kitchen, with its own fireplace for cooking. Other rooms included a small chapel and the private rooms of the lord and his family. This accommodation was probably used by visitors when the castle was busy, because it was more private and safer than the bailey.

Building a Shell Keep

Shell keeps were popular because they could easily be added to the motte and bailey defenses and were fairly cheap to build. However, it was essential that the earth of the motte had been packed down hard enough to take the weight of the heavy stone shell keep. Building such a heavy wall near the edge of the steep motte was dangerous. Sections of the stone wall often fell away during construction, injuring or killing workers.

SQUARE KEEP CASTLES

THE KING AND HIS ATTENDANTS ride toward the new, gleaming square keep that dominates the town. He pulls up his horse and stares proudly at the recently completed plinth, high walls, buttresses, and four lofty corner towers. He jumps off his horse and strides toward the high staircase to inspect inside.

▷ Castle Hedingham in Essex, England, is a wonderful example of a lofty square keep. Note the entrance on the first floor for security and how the windows become wider higher up.

The strength of square keep castles come from sheer size—their thick walls and height. It was essential to build such a high structure on a firm foundation, and the plinth provided this. Made of solid rock, it usually sloped outward. This angle could deflect battering rams, and if a defender dropped a rock from the battlements, it would bounce outward toward the attackers.

Building Square Keep Castles

The size of these towers ("keep" is a much later name) is breathtaking. Base walls 13–16½ ft. (4–5 m) thick that narrow to 8¼ ft. (2.5 m) at the top are usual. The credit for their strength goes to the stonecutters and masons who patiently built up these towers over a long period of time, sometimes up to ten years for one keep. Using hand tools, they sorted and shaped the stones and cemented them together to construct the huge walls, many of which are still standing.

▷ Carpenters were also needed, since wood played a major part in the construction of square keep castles. They used traditional tools, some of which are still used today. This carpenter is using an "auger," the medieval equivalent of a drill.

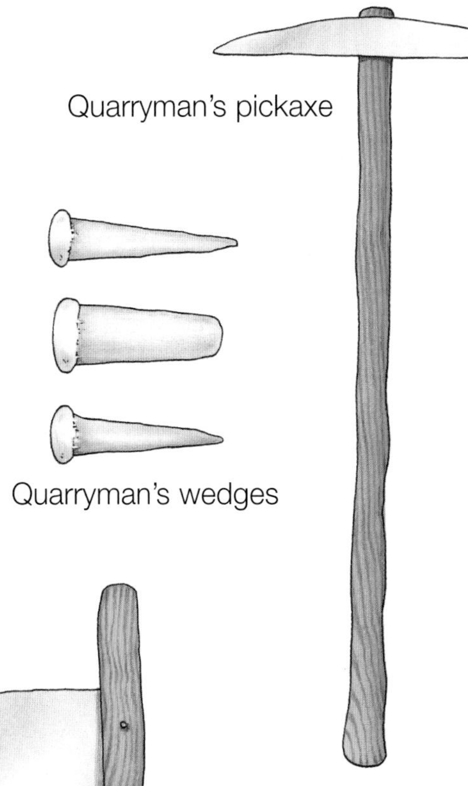

Quarryman's pickaxe

Quarryman's wedges

Stonecutter's saw

INTERIORS

The interiors of these huge towers were impressive—there was a basement and usually three upper floors. How was this achieved? With a vast amount of wood! Whole tree trunks were cut down, trimmed, and then squared into massive beams called *binding joists*. These were inserted into large holes in the walls. These joists became the ceiling of the floor below.

◁ The higher up in a tower, the wider the windows. This meant that rooms near the top were light and airy. Note the thickness of the wall.

Latrines and Wall Chambers

One advantage of thick walls was that it was possible to build small rooms inside them. Known as "privy chambers," they were used by the more important members of the household. Many were equipped with their own toilets, or "garderobes." These were well built, consisting of a wooden or stone seat above an extended chute that opened out halfway down the keep wall above the ditch or in the bailey. Hopefully, the waste was removed regularly.

At right angles to the main joists, bridging joists were laid, which allowed the floorboards to be nailed down. All floors were served by a spiral staircase—a great space saver—usually in one or two of the corner towers. The keep was finished off with pitched or angled roofs, again built with massive oak beams. They were lined with lead to allow the rainwater to run off.

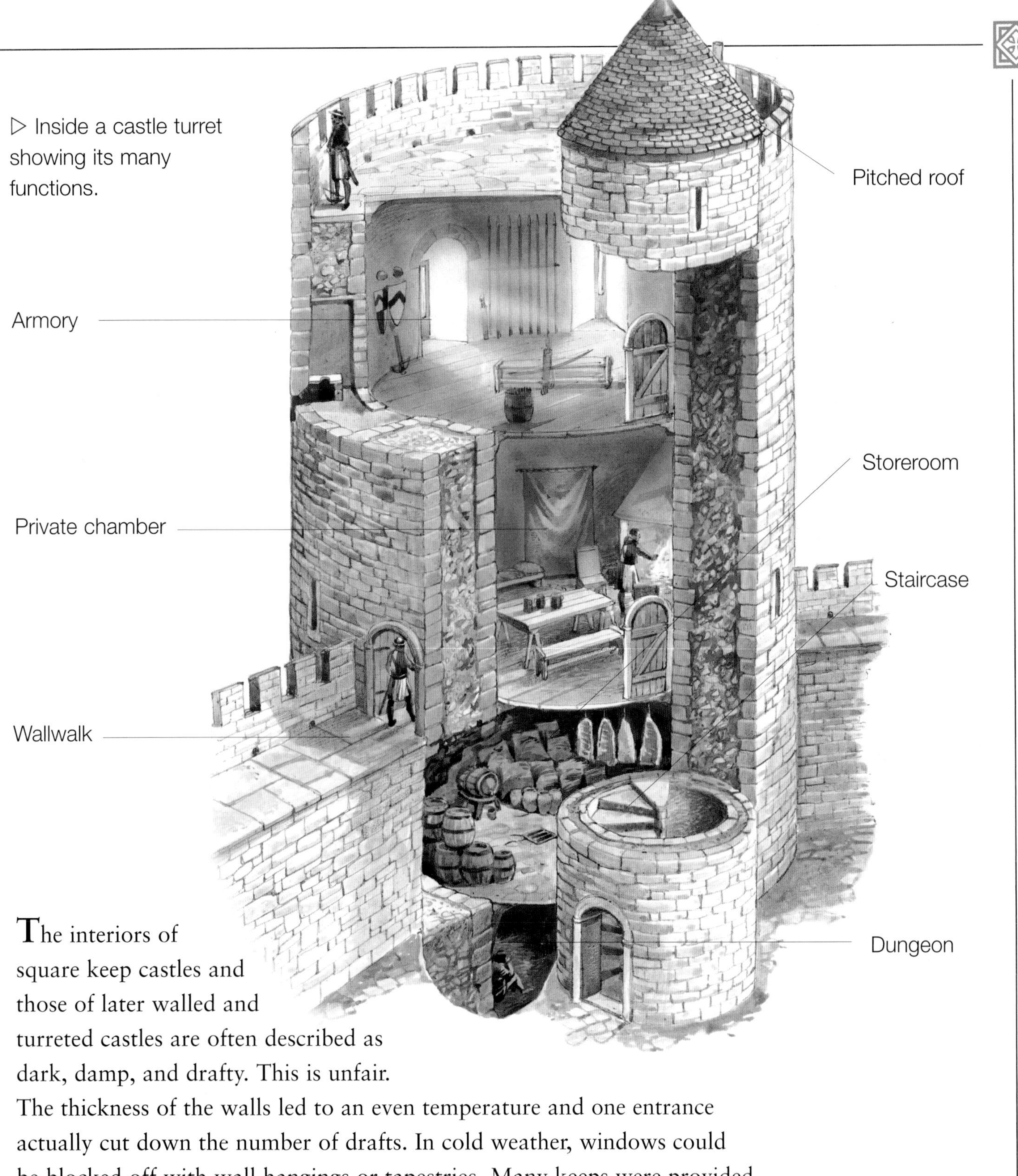

▷ Inside a castle turret showing its many functions.

The interiors of square keep castles and those of later walled and turreted castles are often described as dark, damp, and drafty. This is unfair. The thickness of the walls led to an even temperature and one entrance actually cut down the number of drafts. In cold weather, windows could be blocked off with wall hangings or tapestries. Many keeps were provided with fireplaces to warm the rooms. The cellars were spacious and cool, which helped to preserve food.

DEFENSE INTO ATTACK

THE MASON PUTS THE FINISHING touches to an arrow loop in the southwest tower of the castle, assisted by his apprentice. The construction of the castle has taken almost ten years. He is very proud of the work he has done, and has been training his apprentice carefully. One day he, too, will become a mason and perhaps work on a great castle like this.

During the late twelfth and early thirteenth centuries, great changes in castle design took place. Turrets now projected out from the walls, so there was a clear line of fire along to the next turret. If defenders could fire out of their turrets at the same time, they could trap attackers at the base of the wall—known as *crossfire*. It was much safer than leaning out of an embrasure to fire.

The Crossbow

The crossbow fired bolts (or "quarrels") that were much heavier than arrows, and consisted of a bow fixed across a wooden stock. Some bow-cords were so strong that they had to be drawn back by a winch, and the bow was steadied with a stirrup. The crossbow was such a lethal weapon that the Pope tried to ban it (unsuccessfully) in 1139.

The turrets were called "flanking towers," and they were so effective that after 1200, castles became simple enclosures of turreted curtain walls peppered with hundreds of arrow loops. During an attack, these castles could produce a frightening amount of deadly crossfire from towers, walls, battlements, loops, and firing platforms. A castle's strength now lay in its turrets and walls.

◁ Two flanking towers joined by a somewhat crumbled curtain wall at Carew Castle in Wales. The far tower has an angled, strengthened base.

WHY DID CASTLES CHANGE?

What had brought about this dramatic change in castle development? We have already encountered the answer—the crossbow. During the series of wars between Europe and the Turkish Empire, known in the West as the Crusades, the crossbow was a very important weapon. When soldiers returned from the fighting, many were skilled in either using the weapon or controlling large numbers of crossbowmen. Turned against a castle, they could have a devastating effect.

△ The crossbow in action at the siege of Wartburg. Notice the towers and wallwalk, and the wooden shutter that is protecting one of the defenders.

△ Richard the Lionheart leaving France for the Third Crusade in 1190. He was killed in 1199 by a crossbow bolt fired from a castle.

Crossbowmen placed in siege towers could easily destroy armored soldiers on the wallwalks, whereas flanking towers provided protection and enabled a castle to become an active fighting stronghold. Crossbowmen at the right height could hit the siege towers, attack soldiers approaching the walls, and fire at "sappers," who tried to dig under the curtain wall and bring it down. This threat of "undermining," as it was called, was a major problem for castle defenders.

Build Your Own Castle

Step 1: The Turrets

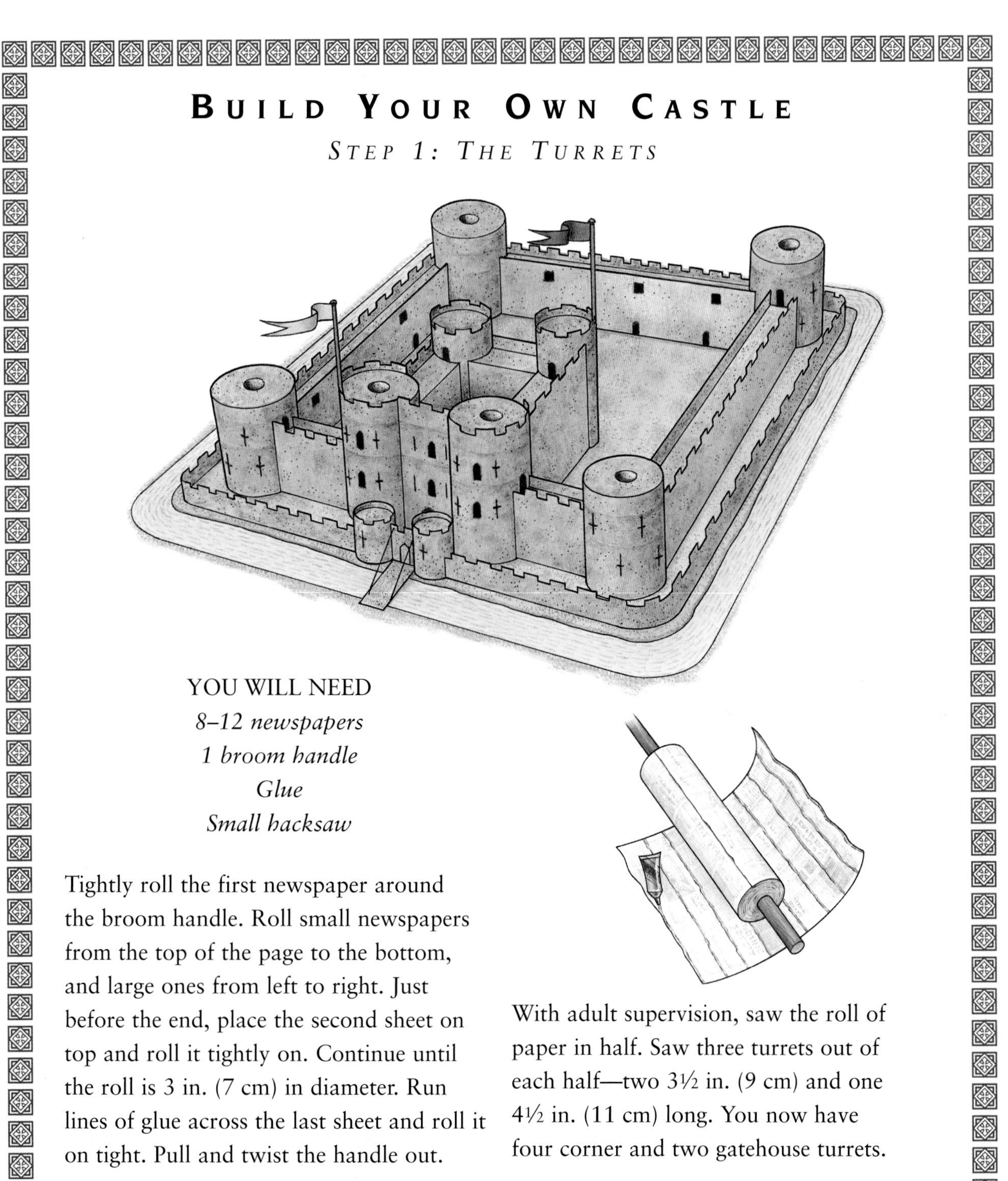

YOU WILL NEED

8–12 newspapers
1 broom handle
Glue
Small hacksaw

Tightly roll the first newspaper around the broom handle. Roll small newspapers from the top of the page to the bottom, and large ones from left to right. Just before the end, place the second sheet on top and roll it tightly on. Continue until the roll is 3 in. (7 cm) in diameter. Run lines of glue across the last sheet and roll it on tight. Pull and twist the handle out.

With adult supervision, saw the roll of paper in half. Saw three turrets out of each half—two 3½ in. (9 cm) and one 4½ in. (11 cm) long. You now have four corner and two gatehouse turrets.

Concentric Castles

THE CROSSBOWMEN on the inner walls of the castle fire over the heads of their comrades on the outer walls at the enemy below. They are confident of victory, because they know that if the attackers get through the outer gate they will be trapped in the narrow passageway between the two sets of walls, which will become a "killing ground."

△ Built after 1205, Krak des Chevaliers has been called the "finest and most romantic castle in the world."

In the thirteenth century, castle engineers took a huge leap forward with the building of "concentric" castles. The first early example to survive is Qalaat-el-Hosn in Syria—known in the West as Krak des Chevaliers – built by the Knights of St. John during the Crusades. Concentric castles have two circuits of walls and flanking towers, but the inner circuit is higher than the outer.

The Engineers

Who the master mason of Krak was is not known. The names of most of the engineers (master masons were both architects and builders) are long forgotten. A small number have been remembered—masons who worked all over Europe, such as Master James of St. George, from Savoy (now part of Switzerland). Henry Yevelle (shown here) designed part of Canterbury Cathedral in England, and is also known to have been consulted about castle building.

Build Your Own Castle

Step 2: Turrets and Walls

YOU WILL NEED

Cardboard (from cereal packages)

Gummed paper tape

2 cardboard tubes (from kitchen towels)

1 large flat piece of cardboard box

(20 in. (50 cm) square)

Glue the four 3½ in. (9 cm) towers 1¼ in. (3 cm) in from the corners of the base. Glue the two large gate towers in the middle, 1¼ in. (3 cm) apart. When they are dry, cut your walls out of cardboard and tape them in between the turrets. Use a lot of tape, since this can be painted over. Tape the kitchen tubes behind the two gate towers, and then cut out and tape the inner walls of the gatehouse.

Building the Castle

As King Edward's ship enters the Menai Strait, with the Island of Anglesey to his right, he sees the towers of his new castle, Beaumaris. He smiles approvingly at the hive of activity. Thousands of men are shaping blocks of stone, winching up baskets of rubble to the working levels. Dozens of roof beams are being slotted into position.

Some of the greatest concentric castles are found in North Wales, built by Edward I against the Welsh, but they exist all across Europe. They were very simple. Like Krak des Chevaliers, an inner space, or ward, was protected by two huge D-shaped gateways, high walls, and massive corner drum towers. This was surrounded by a lower wall with towers, creating a narrower outer ward.

The master mason at Beaumaris was James of St. George. Because it was such an expensive project, he had to send regular reports to Edward, such as this one, dated May 12, 1296: "Some of the castle stands 28 feet high ... We have begun 10 of the outer and four of the inner towers. Four gates have been hung ... and each gateway is to have three portculises ... so much we have done."

Although it was begun in 1295, Beaumaris was such a huge project it was never properly finished. A description from 1341 tells us: "gatehouses and towers are incomplete and some are roofless." Records show that up to 400 masons, 2,000 laborers, 30 smiths, and 200 carters worked on Beaumaris during the summer, which was the main building season.

THE WORKFORCE

Building a stone castle was a difficult project, one that required money, skilled men, laborers, raw materials, and good organization. The work needed an army of men. Craftsmen and laborers were recruited from all over, often press-ganged and marched to the site by troops. The records show that mounted soldiers were paid 7½ pence a day for: "guarding the workmen ... lest they flee on the way."

△ Guilds were formed to protect their members, maintain standards of work, and stop competition. Most skilled trades in medieval times had a guild. On this seal of the Carpenter's Guild in Antwerp, you can see some of their tools.

◁ A tower under construction. Note the wooden scaffolding and the winch lifting the stone block.

Apart from carpenters and masons, there was a huge variety of workmen. Quarriers were needed to provide local stone and carters to carry it to the site. The miners started by digging the huge ditches and later, trenches for the wall foundations. Limeburners made the mortar to cement the blocks. Plumbers and tilers were needed for roofing, and smiths helped with the making of nails, doors, locks, and portcullises.

◁ The powerful walls and towers of Lincoln Castle. It was built so solidly that it successfully beat off a siege in 1216–17.

At the same time, wooden lodges and huts were built to house and shelter the workers. Some were used as workshops, especially during bad weather. Others were used for sleeping. The scene would have resembled a busy wooden city, "with so many smiths, carpenters, and other workmen, working so hard with bustle and noise that a man could hardly hear the one next to him speak."

Build Your Own Castle

Step 3: Finishing Touches

YOU WILL NEED
Thin cardboard
Paints
Paintbrush
An old toothbrush
Colored paper
Cocktail sticks

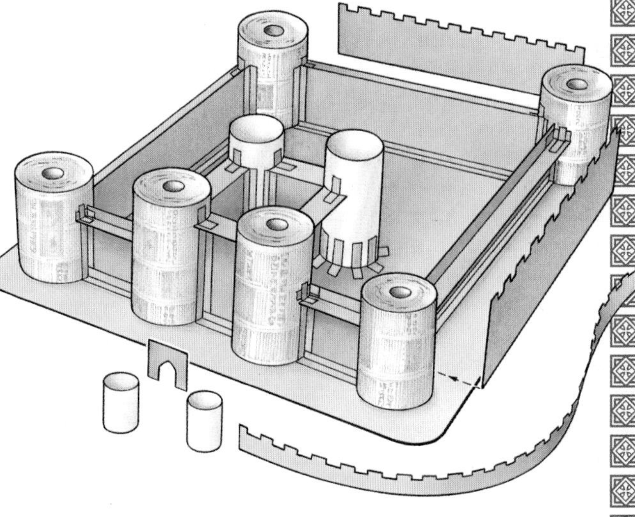

Cut out your battlements and tape them to the front of the walls and turrets. Add a crenellated outer wall—lower than the inside wall—with cardboard and tape. Cut out a cardboard rectangle for the drawbridge. Paint all the walls and towers gray. When they are dry, "spray" the walls other colors like dark brown, light brown, and red, using the toothbrush, to give a stone effect. Paint the wards a sandy color, and the arrow loops black. With the cocktail sticks and colored paper, decorate the towers with flags.

BUILDING WITH STONE

Before the building could actually start, the raw materials had to be transported to the site. Most of them were very bulky and needed large numbers of carters. If the castle site was near a harbour or on a river, heavy material could be carried by ship or barge. The most important materials were stone, rubble, timber, ropes, limestone, sand, iron, and lead. Officials made sure the correct amounts were delivered.

△ A mason "dresses" a block of stone with a chisel, under a very watchful eye.

Castle walls were built with a fairly rough masonry core—usually flint—covered in rough or dressed stone. Important buildings, such as keeps, towers, and barbicans, often had "dressed" stone cemented onto the front, and most certainly on corners and openings like archways. Dressed stone (called *ashlar*) had to be carefully cut, chiseled, and hammered smooth, and so was very expensive. Because of this, it was used sparingly, unless a lot of money was available, as in the case of a royal castle.

MORTAR

To cement the masonry together, the masons used mortar. They usually mixed it in a large trough or pit, by adding sand to quicklime (burned limestone). The lime absorbed carbon dioxide in the air, making a hard calcium carbonate. This meant the mortar actually gripped the stone instead of splitting off.

Mortar-maker's hoe

◁ These tools were used to build a Spanish monastery but are exactly the same as those used on castles. Notice how the rope passes through the block and tackle to make hauling easier, and the "pincer" tools; a similar tool can be seen in use in the illustration of the tower on page 32.

It was difficult and dangerous to lift stone onto high building positions. The masons used a variety of technologies to achieve this. We know that scaffolding was used, both from contemporary pictures and because the holes for the beams—called *putlogs*—can still be seen in castle walls. Winches or hoists to haul the stone were made of wooden uprights and crossbeams, thick ropes, and block and tackle. Blocks, gripped with iron tongs, were pulled up by human muscle power.

▷ The flint core and ashlar of the keep at Old Sarum. Note the angled base. Most of the dressed stone from this castle was robbed and used in buildings in nearby Salisbury.

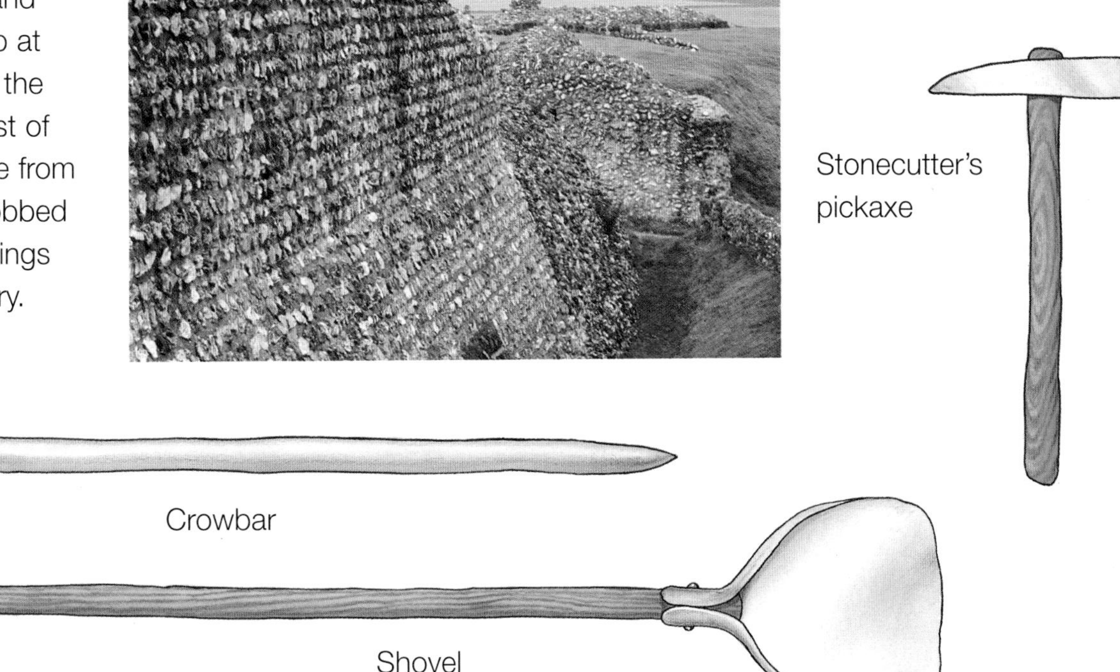

BUILDING WITH WOOD

It is easy to lose sight of the importance of timber in stone castles. When Edward III decided to enlarge Windsor Castle—his birthplace—3,994 oaks were felled at "a certain wood at Cagham" in 1354, and again, over 2,000 more in 1361. The king's clerk, Richard of Rotheley, was instructed "to provide masons and carpenters for our works, wherever they can be found."

◁ King Edward III changed the interior of the shell keep at Windsor Castle from a fortress building to a comfortable family home.

The timber was used for the chapel, and the roof and fittings of the new hall, new bath houses, the queen's "dancing chamber," pens for the king's falcons, and a new kitchen and gatehouse. Unsurprisingly, except for the king, "hardly anyone could get a good carpenter." The total cost was a massive $100,000 (£50,000), the equivalent of $200 million (£100 million) today!

△ Two sawyers at work. The unpleasant position of the lower man covered in sawdust compared with that of the man above him gave rise to the expression *top notch*, meaning in a better or favored position.

THE PIPE ROLLS

▷ A French castle under construction. Look at all the different skills and equipment used. The nobleman on horseback is probably checking up to see that his money is being properly spent.

We know a lot about the types of jobs carried out during castle building from some very old and important documents—the "Pipe Rolls," or great "Rolls of the Exchequer." They were begun before 1130 by Henry I and continued right up until 1832, showing all the money spent in the name of the king. Year after year, castle building was easily the heaviest item of royal spending.

WHY "PIPE ROLLS"?

The Pipe Rolls are the oldest and longest series of public records known in England, and provide useful information about medieval finances, such as the cost of castle building and revenue from feudal dues. Henry I introduced them to ensure that the kingdom ran smoothly during his long absences abroad, although only one example survives from his reign. The name *Pipe Rolls* probably comes from the fact that when the sheepskin rolls were folded, they looked like a stack of pipes.

It is because of the very accurate bookkeeping that we know so much about the details of castle building. They tell us that progress was dependent on the weather, that two-thirds of the cost of a castle went to wages, and that the number of workers at any one time was huge. Some men were paid according to quantity of work, and others according to the number of hours they had worked each day. Skilled men were not paid by the day, but only once every month.

Castles around the World

Japanese Castles

A LONE MOUNTED SAMURAI warrior raced toward the castle. With his lord's colored pennant fluttering on his back, he is recognized by the guards and allowed across the moat bridge. Heavy double doors swing open and the Samurai makes his way along the narrow corridors. Ushered into his lord's presence, he bows and breathlessly informs him that the rival Daimyo's army marches on the castle.

◁ A Samurai warrior, known as a *bushi*, pays homage (loyalty) to his Daimyo. Samurai without a lord were known as *ronin*.

By the sixteenth century, Japanese castles were a cross between a military base that could withstand muskets and small cannons, and a comfortable home. They were built on a strong, skillfully stacked base of massive stones, sharply angled against battering rams. This was crowned by a huge timber keep with thick mud and plaster walls. The keep could have as many as five floors, each smaller than the one below it and with its own tiled roof. Surrounded by moats and drawbridges, the entrances were zigzagged to confuse attackers.

▷ A typical Japanese castle with a Samurai guard. Such castles had an outer wall and moat several miles long, enclosing a second wall and moat. Behind this were watchtowers and high earth banks.

The Daimyo System

In Japan, after the twelfth century, a system similar to European feudalism developed. At the top was the *Shogun*, or "great general." Below him were the landowning warriors, similar to European barons called *Daimyos*, or "great names." These men were very powerful and with their armies of Samurai warriors, often went to war with each other. After 1603, the Shogun forbade any Daimyo from having more than one castle.

AN INDIAN CASTLE

The Emperor Akbar, ruler of the Mughal Empire in sixteenth-century India, decided his capital Agra, in northern India, should have a fortress to be proud of. He ordered the demolition of the old brick fort of Sikander Lodi, which was "ugly and ruinous" and laid the foundation of the new fort in the spring of 1565.

The chief engineer who planned and built the fort was Qasim Khan—a talented man who as well as being superintendent of public works was also chief of the navy. He designed the fort to be semicircular in shape, with an enclosing wall of 1½ miles (2.4 kilometers). The foundations were dug very deep to accommodate the high walls, which are over a yard thick. The stones of the walls were laid closely together and fastened with mortar and iron rings.

△ The Amar Singh gate of the Red Fort, Agra. Note how the narrow embrasures on the wallwalk gave extra protection to the defenders.

Building the Red Fort

The building of the fort was a huge undertaking. Over 3,000 masons and 8,000 laborers were daily employed on the work. Qasim Khan took over 1,000 bullock carts to bring red sandstone in from nearby quarries.

"The continued noise from the construction meant Akbar's family and court had to move to Nagarchain 'The Abode of Delight' 7½ miles away to afford them relief."

△ Construction of the Delhi gate, with its octagonal towers. Because the Red Fort was built next to the Yamuna river, much stone was carried to the site by hundreds of boats.

The fort was built was very bright red sandstone which gives rise to its name, the "red fort." Its walls were very well-designed to withstand sieges, with battlements, embrasures, machicolations, and arrow loops. It has two gateways, the Amar Singh gate and the Delhi gate, with bridges crossing the surrounding moat, 7½ yd. (7 m) wide and 4½ yd. (4 m) deep. The fort took 15 years to build, although the Emperor moved in in 1569 and began building his splendid domed palaces.

The Castle Legacy

A GROUP OF WORKMEN hammer at the tops of the flint wall, and blocks and mortar come crashing down. Bit by bit, the castle is being demolished and sold off as building stone. Tomorrow an engineer will arrive to destroy the huge curtain wall. His miners will dig shafts underneath the foundations and insert wooden props. These will be set on fire and sections of wall will fall over.

The Fortress Lived On

At first, castles adapted well to cannons and the Italians developed artillery forts. Then the work of Frenchman Sebastian de Vauban in the late Seventeenth century really influenced the next 150 years of fortress building. His principle was to keep an attacking army as far away as possible by complicated outworks, moats, and pointed bastions. His network of fortresses saved France in the wars against Britain and Holland during 1702–13. Although her armies were defeated, the line of French fortresses wore out the British and Dutch advances. Paris never fell.

Although this was the fate of many castles in the seventeenth and eighteenth centuries, it wasn't only gunpowder and cannons that ended the importance of the castle. The decline of the feudal system (which provided cheap manual labor) and lack of money to maintain castles made the barons look for other ways to defend their local power. Castles were not needed any more. Some served as local jails, but little else.

Castles were given a new lease of life during the English Civil War between 1642 and 1653. Many castles were under siege and some held out bravely for many weeks. In fact, there were more sieges than pitched battles during this war. When Oliver Cromwell and his forces recaptured them, he had them "slighted"—the battlements and gateways destroyed—so they couldn't be used against him again. However, castles, even in such a ruinous state, still hold a fascination for many.

▽ Corfe Castle in Dorset, with its huge keep a broken and battered silhouette. It was the scene of a struggle between the King's forces and Parliament during the English Civil War.

TIMELINE

950	995	c. 1000	c. 1035
Doue-la-Fontaine begun.	Langeais begun. Both in northwest France. Earliest surviving castles.	Rouen, Bayeux, and Ivry Castles built in Normandy.	Tilliers, Falaise, Cherbourg, and Caen Castles built in Normandy.

c. 1100	1127	1197–8	c. 1205
Shell keeps constructed on existing mottes. Stone curtain walls begin to replace wooden palisade fences.	Rochester Castle—a massive square keep castle—begun in England.	Richard the Lionheart constructs Chateau Gaillard in France.	The huge castle of Krak des Chevaliers in Syria built. This was to have a great influence on European castle building.

1350	14TH CENTURY	c. 1450	c. 1500
Edward III begins rebuilding Windsor Castle in England.	Gunpowder used against castles.	Great age of "Daimyo" castles in Japan.	Artillery forts first built by the Italians.

1066

Norman invasion of England.

William the Conquerer builds a simple wooden defense inside the Roman fort in Pevensey, England.

Hastings Castle built.

c. 1070

Square keep castles begun in London (the White Tower) and Colchester in England.

1095

First Crusade begins. Christians from Europe try to take the Holy Land.

13th Century

Castle walls strengthened by flanking towers or turrets.

1277

King Edward I invades Wales.

1283

Edward I begins the concentric castle in Harlech, and the massive turreted castles in Conwy and Carnarvon in Wales.

1293

Beaumaris Castle in Anglesey, Wales, begun.

1565

The Red Fort in Agra in northern India begun.

1642–55

The English Civil Wars. Castles play an important part.

1698–1708

Neuf Brisach built in France. Designed by the pioneer of castle design, Sebastien de Vauban.

GLOSSARY

Anglo-Saxon Chronicle An early record of events in England until 1154.

Bailey The large enclosed courtyard of an early castle.

Barbican An extra defensive gateway built in front of the entrance.

Barons Powerful men given large amounts of land by the king in return for support and soldiers.

Bastion tower Turret or projection from a wall to increase its protection.

Buttery Storeroom where drink (usually ale), butter, cheese, and bread were stored and given out.

Buttresses Stonework built against a wall to give it extra strength.

Cavalry Soldiers who fight on horseback.

Concentric castle Castle with two circuits of walls and flanking towers, with the inner wall higher than the outer.

Corbel Stone or wooden projection from a wall to support a beam or other weight.

Crenellate To build battlements on a castle wall.

Crusades Religious wars between Christianity and Islam that began in the eleventh century.

Curtain wall Continuous wall enclosing a bailey or a wall.

Dressed stone Also known as ashlar. Stone that was carefully cut and smoothed by masons.

Embrasure Gap in a parapet for soldiers to fire through.

Flanking towers Towers projecting from the walls of a castle.

Flint Hard, sharp rock found in layers in chalk.

Garderobe Toilet built into the outside wall of a castle tower.

Joists Large wooden beams supporting the floors of castle towers.

Knight A mounted warrior who received land—a manor—if he promised to fight for the baron.

Loop Vertical slot in a wall to fire out of—thus, "arrow loop."

Medieval The Middle Ages, between A.D. 500 and 1500.

Mortar Mixture of lime, sand, and water that holds stone together.

Motte A tall mound of earth on which early castles were built.

Posthole Usually a circular hole built in the ground to receive a timber post.

Press-ganged Forced into doing something you do not want to.

Ringwork Circular ditches and banks built to protect early castles.

Sappers Men who undermined the foundations of a castle's curtain walls in an attempt to make them collapse.

Shell keep Hollow, stone, circular wall built around the lip of the motte.

FURTHER INFORMATION

BOOKS

Castles and Forts (Kingfisher Knowledge) by Simon Adams
(Kingfisher, 2003)

Knights and Castles (Magic Tree House Research Guide)
by Will and Ma Osborne (Random House Books for Young Readers, 2000)

Medieval Life by Andrew Langley
(Dorling Kindersley, 2002)

Royal Castle (Inside Story) by John Farndon
(Penguin UK, 1999)

Scottish Castles Through History by Richard Dargie
(Hodder Wayland, 1998)

The Best-Ever Book of Castles by Philip Steele
(Kingfisher, 1999)

Web Sites
Due to the changing nature of Internet links, PowerKids Press has developed an online list of Web sites related to the subject of this book. This site is regularly updated. Please use this link to access this list:
www.powerkidslinks.com/aoc/built

INDEX